故園畫憶

庚寅中秋

韩启德题

故园画忆系列

Memory of the Old Home in Sketches

豫南印象

Impression of South Henan

张可松　绘画 撰文

Sketches & Notes by Zhang Kesong

學苑出版社

Academy Press

图书在版编目（CIP）数据

豫南印象 / 张可松绘画、撰文. — 北京：学苑出版社，2015.6
（故园画忆系列）
ISBN 978-7-5077-4786-7

Ⅰ. ①豫… Ⅱ. ①张… Ⅲ. ①建筑画—钢笔画—作品集—中国—现代
Ⅳ. ①TU-881.2

中国版本图书馆CIP数据核字（2015）第129150号

出 版 人：孟　白
责任编辑：周　鼎
编　　辑：李点点
出版发行：学苑出版社
社　　址：北京市丰台区南方庄2号院1号楼
邮政编码：100079
网　　址：www.book001.com
电子信箱：xueyuanpress@163.com
销售电话：010-67601001（销售部）、67603091（总编室）
经　　销：全国新华书店
印 刷 厂：北京信彩瑞禾印刷厂
开本尺寸：889 x 1194　1/24
印　　张：6.75
字　　数：150千字
图　　幅：144幅
版　　次：2015年6月北京第1版
印　　次：2015年6月北京第1次印刷
定　　价：45.00元

目　录

Contents

Zhumadian City

Xinyang City

自　序

河南省地处中原，简称豫，从夏代到北宋，先后有20个朝代建都或迁都于此，是中华民族最为重要的发祥地和发源地之一。“豫南”指河南的南部地区，包括南阳市、驻马店市、信阳市及其所辖区域。

南阳市位于河南省西南部，古称宛，与湖北省、陕西省接壤，因地处伏牛山以南，汉水之北而得名。历史上，南阳是古丝绸之路的源头之一，战国时期，南阳是著名的冶铁中心，后为秦朝三十六郡之一的南阳郡治所所在地。东汉时期曾作为陪都，故又有“南都”“帝乡”之称。

驻马市店别称“天中”“驿城”“驿都”，地处淮河流域，因“旧为遂平至明港间驿马驻所”而得名。在古代为“豫州”“蔡州”“汝宁府”等州治府治，至今仍留有“蔡国故城”“汝南古城”“南海禅寺”等遗址。驻马店是中华民族姓氏文化发源地之一，周氏、蔡氏、金氏、江氏家族的发祥地；也是中国车舆文化之乡、中国冶铁铸剑文化之乡、中国嫘祖文化之乡。

信阳市位于河南省南部，在唐代时简称“申州”，故又称“申城”。它是江淮河汉间的战略要地，也是“鄂豫皖”区域性中心城市。自古以来人杰地灵、英雄辈出。早在8000多年前，境内淮河两岸就出现了相当规模的原始农业，从东到西分布有裴李岗文化、龙山文化和屈家岭文化遗址多处。历经春秋、战国之后，中原文化与楚文化在此交融共生，酿就了信阳独具特色的淮上文化风情。信阳地处江淮之间，是历代中原人南迁的始发地和集散地，享有“中原第一侨乡”之美称。

《豫南印象》是以以南阳市、驻马店市、信阳市及其所辖区域所构成的人文历史故事、遗址公园、风景名胜为素材，利用艺术的手法速写出126幅图画，并配以简短的文字叙述，旨在对豫南地区人文景观、文化传统进行记录，希望能在社会中引起读者的共鸣，进而传承发扬中原文化。

《豫南印象》的结集成书非吾一人之功劳，河南的厚重文化亦非本人能表述明了，能够用绘画之技记录下文化的影子，着实是我的福分，艺无止境，望大家指正。

张可松

2014年11月

Preface

Henan Province is located in central China and often called "Yu" for short. From the Xia Dynasty (2070 BC-1600 BC) to the Northern Song Dynasty (960-1127), 20 dynasties have established or moved their capitals here successively. It is one of the most important birthplaces and cradles of the Chinese nation.South Henan refers to the cities of Nanyang, Zhumadian and Xinyang.

Historically, Nanyang was known for being one of the starting points of the ancient Silk Road. The historical origins of many ancient stories can be found here and numerous renowned persons were born in Yu. Over 2,000 years ago, this region became a famous center for iron smelting.

Zhumadian is one of the birthplaces of numerous Chinese family names, including the Zhou, Cai, Jin and Jiang families. It is also the birthplace of many Chinese festivals and legends, like Pan Gu creating the sky and the earth.

Since ancient times, Xinyang has been reputed as a fair place where outstanding people and heroes are produced. There are many scenic spots, like Jigong Mountain. As early as 8,000 years ago, large-scale primitive agriculture came to both banks of the Huaihe River. When emigrating to the south, people from central China first gathered here and then moved on together, so Xinyang was regarded as the "first hometown of emigrated Chinese people to central China."

This book will document the regional characteristics of Southern Henan. It presents relics, historical stories, landscapes and scenic spots in Nanyang, Zhumadian and Xinyang through 133 sketches with captions. The aim is to record the human landscapes and cultural traditions in Southern Henan. I hope that this book helps readers to resonate with the area so as to carry on and disseminate our splendid Chinese culture.

Zhang Kesong
November, 2014

南阳市
Nanyang City

卧龙岗

位于南阳市城西卧龙区，初建于魏晋，盛于唐宋。是三国时期杰出的政治家、军事家诸葛亮躬耕隐居地，汉昭烈皇帝（刘备）三顾茅庐处，“三分天下”的策源地，也是历代祭祀诸葛亮的地方。久负盛名，被誉为“天下第一岗”。

Wolong Hill

Located in the Wolong District in the west of Nanyang city, Wolong Hill is the place where Zhuge Liang, an outstanding politician and strategist in the Three Kingdoms Period (220-280), lived on a farm as a recluse. Since then it has become a place where people offer sacrifices to him.

水帘洞

位于桐柏县城西。水帘洞距地高约 20 余米，一弘山泉自崖巅倾泻下来，像一条水晶挂帘悬在空中，将峭崖上部的一座天然石窟遮掩在幕后。沿石壁有阶绨和铁链可攀援而上，进入洞中，洞内有泥塑猕猴一尊。水帘洞的水，清纯甘冽。夏季凉爽，为避暑胜地。

Shuilian Cave (Water Curtain Cave)

Located to the west of Tongbai County, Water Curtain Cave is about 20 meters above ground. Next to a stone wall, there are stairs and iron chains for tourists to climb up into the cave, in which there is a clay sculpture of a macaque.

水帘寺

位于桐柏县城西，因紧挨水帘洞而得名。始建于宋元佑三年（1088 年）以前，明清都进行过修建，原有山门一间，中殿、后殿各三间，另有陪房十间。寺内墙上嵌有历代文人游客书写的诗文和题记。水帘洞先被道家定为“天下第四十一福地”。清乾隆四十九年（1784 年）以后，桐柏山佛教昌盛，自成白云山系。

Shuilian Temple (Water Curtain Temple)

Located to the west of Tongbai County, Water Curtain Temple was built before 1088 and became a holy site for Buddhism in 1784. Going through the gate, there are middle halls, back halls, and surrounding rooms within the yard. Poems and the calligraphy of ancient scholars and tourists inscribed on the walls inside the temple are still discernible.

菩提寺

位于唐河县城的老城区解放路北侧。寺内藏经楼重檐歇山顶结构，斗拱彩绘，华丽又不失古朴清雅。两侧的建筑对称分列于左右，雕绘也以古朴清雅而见长。殿房、禅院、僧房和藏经楼下种植有奇花异卉。寺后山坳里泉水穿漕入院，茵茵送爽。

Bodhi Temple

Located north of Jiefang Road in the old town of Tanghe County, the Bodhi Temple's Depository of Buddhist Sutras has a gable and hipped roof with double eaves and is flanked by symmetric buildings. On the walls of these two buildings are different kinds of inscriptions and paintings. Exotic and rare flowers are planted here, irrigated by the spring water flowing in from the back of the temple.

清真西寺

位于邓州市城区内，始建于明朝中叶，重建于清道光年间（1821～1850年）。全寺占地面积1000平方米，礼拜大殿为古典式。本坊教民约8000多人，均属回族，遵行格迪目。有阿訇五人、满拉九人、学董及乡老12人。

West Mosque

Located in the seat of Dengzhou City, the West Mosque was built in the mid-Ming Dynasty (1368-1644). The entire mosque covers an area of 1,000 square meters and worship ceremonies are conducted in the classic grand hall.

普化讲寺

位于唐河县城南的黑龙镇发山之顶，为始建于元代的佛教寺庙，明清及民国期间，寺庙屡毁屡建。1940 年，该寺改名为普化讲寺。

Puhuajiang Temple on Fashan Mountain

Located on top of Fashan Mountain in Heilong Town, to the south of Tanghe County, Puhuajiang Temple has a history of over 700 years. It has been destroyed and rebuilt many times throughout its history.

南阳府衙

通常称为府衙或知府衙门，故址在旧城内西南隅，即今南阳城区民主街西端北侧，是元明清三代南阳知府的官署。

Nanyang Government Office

Normally also called the office of a magistrate, Nanyang Government Office was the headquarters of three administrations under magistrates in the Yuan(1206~1368), Ming(1368~1644) and Qing(1644~1911) Dynasties in Nanyang. This antique building is located in the southwest corner of the old town, on the northwest end of Minzhu Street in Nanyang City.

玄妙观

位于南阳市区建设路中段。著名道观，始建于元至元年间（1264～1294年），明正统年间（1436～1449年）重修，现存建筑300余间，中线上有四神殿、正殿、三清殿、二神殿和藏经阁、老君堂等。

Xuanmiao Taoist Temple

Located at the middle section of Jianshe Road in Nanyang City, Xuanmiao Taoist Temple was built between 1264 and 1294 in the Yuan Dynasty. Nowadays, over 300 rooms are left, including the main hall and the "Depository of Buddhist Texts" along the central hall.

医圣祠

位于南阳市城东温凉河畔，是东汉时期伟大的医学家、被尊为“医圣”的张仲景的墓祠纪念地。医圣祠坐北朝南，占地一万多平方米。其始建年代不详，后世多次扩建和修葺。现大门为仿汉建筑，大门内十米处为张仲景墓。建筑群整体为汉代艺术风格，博大雄浑，巍峨壮观。

Medical Sage Temple

Located at the Wenliang River in eastern Nanyang, the Medical Sage Temple is a memorial for Zhang Zhongjing, a great doctor in the East Han Dynasty (25-220).Its gate faces south and its date of construction is unknown.

文笔峰塔

位于唐河县城东南角岗坡上，古人为培植唐河文风而建。塔呈八棱锥形，古砖砌成，高约 30 米，九级。底层周长 17 米。外壁有题记和壁画，与城内文庙遥遥相望。塔形酷似一支饱蘸浓墨的神来之笔，挺拔俊秀、巍然屹立。与城北的泗洲寺塔遥相呼应，素有“一城担二塔，二塔抬一城”之说。

Wenbifeng Tower

Located on the hillside southeast of Tanghe County, Wenbifeng Tower is an octahedral-pyramid-shaped building made of bricks. The 30-meter tower is divided into 9 floors. The perimeter of the ground floor is 17 meters. It was decorated with calligraphy and paintings on the outside.

炼真宫

位于方城县县城北部，宫城为四边各长 150 米的方形，为土筑，残高五至六米，厚二至三米。据明《重修炼真宫记》碑文记载，此处“乃东汉湖阳公主修真之所也。”湖阳公主为汉光武帝刘秀长姐，夫亡后欲嫁大中大夫宋弘，宋弘为妻守节而不娶，并于此地守真全节。后来葛玄、张三丰等道教名士在此修炼。每逢农历初一、十五，游人香客甚众。

Lianzhen Palace

Located in the north of Fangcheng County, Lianzhen Palace is a square-shaped building with four sides of 150 meters each. Its wall was made of earth and the left wall is 5-6 meter high and 2-3 meter thick. As a holy place for Taoism, it attracts a lot of pilgrims.

佛沟摩崖造像

位于方城县小史店乡的桐柏山余脉中，开凿于北魏至唐代，当地群众称之为“佛沟”。造像分别镌刻在南北两块天然巨石上，造像共 32 龛 138 尊，最大佛像高 1.4 米，最小者 0.2 米。造像形态多变，有较高的艺术、佛学价值。

Statue on the Fogoumo Cliff

Located in the branch of the Tongbai Mountain range in Xiaoshidian Township of Fangcheng County, these statues were carved during the period between the Northern Wei Dynasty (386-534) and the Tang Dynasty (618-907). 138 unique statues are carved into 32 niches on two huge, naturally formed rocks to the south and north respectively.

普严寺

又称普严禅院、大寺。位于方城县大乘山下，始建于唐贞观年间（627～649 年），后世曾进行重建和整修。为河南四大名寺之一，先后有多位禅师在此驻锡传灯。建有山门、中佛殿、大雄宝殿及两庑。门前有两棵千年银杏树，枝叶繁茂，苍劲挺拔。

Puyan Temple

As one of the four most famous temples in Henan, Puyan Temple is also called Puyan Buddhist Yard or the Grand Temple. It is located at the foot of Dacheng Mountain in Fangcheng County and was built during Emperor Zhenguan's reign(627-649)during the Tang Dynasty (618-907). Two maidenhair trees that were planted 1,000 years ago are still thriving in front of the gate.

花洲书院

位于邓州市，始建于宋庆历年间（1041～1048 年），北宋著名政治家、军事家、文学家、教育家、思想家范仲淹任邓州知州期间因百花洲一带环境幽静，景色宜人而创建书院内讲学堂，花洲书院因此而得名。

Huazhou Academy

Located in Dengzhou City, Huazhou Academy has a history of nearly 1,000 years. Fan Zhongyan, a thinker in the Northern Song Dynasty (960-1127), built this academy when he was serving as the governor of Dengzhou because he liked the tranquility in Baihuazhou, which is favorable for studying.

淅川香严寺

又名长寿寺、香严长寿寺、显通禅寺。位于淅川县仓房镇境内，始建于唐代，四面群山环抱，整个地形若莲花状，寺居正中。原有两座禅院，为上寺、下寺。现仅存上寺。

Xiangyan Temple in Xichuan

Located in Cangfang Town of xichuan County, Xiangyan Temple is also known as the Longevity Temple or Xiantong Buddhist Temple. Surrounded by mountains, the temple sits in the center of the lotus-shaped terrain. It was built in the Tang Dynasty (618-907) and was composed of two yards among which only the upper yard is left.

诸葛庐

南阳诸葛庐位于卧龙岗之上，初建于魏晋，盛于唐宋。是三国时期著名的政治家、军事家诸葛亮隐居时的草庐，名胜古迹。

Zhuge's Cottage

Zhuge's Cottage was built during the Wei and Jin dynasties and became famous in the Tang and Song dynasties. Zhuge Liang (181-234), a famous strategist in the Three Kingdoms period (220-280), lived as a recluse in this cottage, which is located on Wolong Hill.

显圣庙戏楼

位于内乡县王店镇西北处，坐南面北，与大殿相对应，通面阔三间，长 7.2 米，进深 7.6 米。戏楼建在一长方形直壁式台基之上，戏台基座高 1.5 米，戏楼通高约 7.5 米。屋内梁枋上刻字说明为元代所建。

Opera House of Xiansheng Temple

The Opera House of Xiansheng Temple is a south-facing building located in the northwest of Wangdian Town of Neixiang County. It is 7.2 meters wide and 7.6 meters long. If the 1.5-meter base is counted, the house is 7.5 meters tall in total. The letters engraved on the beam inside the building indicate that it was built in the Yuan Dynasty (1206-1368).

丹水镇戏楼

丹水古镇历史悠久，自西周时期就名誉中原，现名为“中州古镇”。戏楼建于清代，如仙阁琼台，楼身飞檐重重，饰有彩瓷镶嵌的图案，柱梁上有镂空木雕，玲珑剔透，巧夺天工，雕刻形象惟妙惟肖，显示出极高的建筑艺术造诣。

Opera House of Danshui Town

Danshui became famous in central China in the Western Zhou Dynasty (1046 BC-771 BC) and is called “the Old Town of Zhongzhou” (meaning “central China”).It looks like a richly decorated jade palace with its colorful porcelain walls and hollowed-out wooden works carved on its pillars .

社旗山陕会馆

位于社旗县城中心，始建于清乾隆二十一年（1756 年）。清时这里商业极盛，由旅居该地的山陕富商出资兴建。会馆坐北朝南，总占地一万多平方米。建筑布局采用了中轴对称式的传统风格，有戏楼、拜殿等，既雄伟壮观又典雅有致。有“天下第一会馆”之美誉。

Shanxi and Shaanxi Guild Hall in Sheqi

Located in the center of Sheqi County, the Guild Hall was built in 1756. Rich businessmen from Shanxi and Shaanxi traveling here sponsored its construction during the most prosperous period of the Qing Dynasty (1644-1911). The south-facing building has a traditional architectural style of symmetry and a central axis.

陕西会馆

位于唐河县北源潭乡。建于清乾隆年间（1736～1795年），馆内现存大殿、配殿及东厢房等。大殿为会馆主体建筑，宏伟壮观。上层檐顶用琉璃脊，正脊两侧雕有飞龙，正脊中央有一高达两米的琉璃重檐歇山顶建筑。殿内雕梁画栋，十分精美。

Shaanxi Guild Hall

Located in Beiyuantan Township of Tanghe County, Shaanxi Guild Hall was built during Emperor Qianlong's reign (1736-1795) in the Qing Dynasty (1644-1911).The existing hall is composed of a main hall, side halls, and east surrounding rooms.There are many painted decorations inside the focal building of the main hall.

汲滩陕山会馆

位于邓州市城东 20 千米的汲滩镇中学校院内。是外地旅邓商人聚会议事的地点。也是用来敬神感灵，祈福禳祸，交流信息，接待商人和停放货物的地方。

Shaanxi and Shanxi Guild Hall in Jitan

Located in the middle school of Jitan Town, 20 kilometers east of Dengzhou City, this guild hall was a place for local merchants to get together and have discussions. It was also used for worshiping gods, exchanging messages, receiving merchants, and storing goods.

内乡县衙博物馆

内乡县衙始建于元大德八年（1304 年），重建于清代，占地两万多平方米，厅堂房舍 280 余间。县衙坐北朝南，存房屋 98 间，建筑面积 2704 平方米，有大门面阔三间，大堂面阔五间。是我国封建社会县级政权衙门的实物标本和历史见证，是一座十分珍贵的文史资料库，保存较完好。

Neixiang County Government Office Museum

Built in the 8th year of Emperor Dade's reign in the Yuan Dynasty (1304), the museum was rebuilt during the Qing Dynasty (1644-1911). The south-facing museum is composed of more than 280 halls and houses, among which only 98 are left.

吾离冢遗址

又名“五离冢”“五女冢”，位于邓州市东南两千米处，相传为春秋时期邓侯吾离冢。《春秋左传》记载，吾离是曼姓邓国第十九位邓侯，也是第一位见于正史的邓国国君，有“中兴之君”之盛名，其陵墓是现存唯一的邓氏先祖陵墓，人称“天下邓氏第一陵”。

Wuli's Tomb

Located 2,000 meters southeast of Dengzhou city, it is also known as “Wunv's Tomb”. This place is said to be the tomb of Wuli, king of Deng Zhou in the Spring and Autumn Period (770BC-476BC). It is recorded that Wuli has a family name of Man (a Chinese name) and he was the 19th king of the Deng Kingdom. His tomb is the only one left in the entire Deng family.

古蓼国遗址

位于唐河县湖阳镇，东北依蓼山，山下有蓼王叔安疏洪治水开凿的人工河蓼阳河、蓼阴河，是世界廖氏发祥地。蓼为已姓国，歙项帝后裔。

Acient Liao Kingdom

Located in Huyang Township of Tanghe County, the kingdom was backed by Liaoshan Mountain in the northeast. It was at the foot of the mountain that Liao Shu An, the King of the Liao Kingdom, led flood-control projects, including the digging of artificial channels for the Liaoyang River and Liaoyin River. It is also known as the birthplace of the family name “Liao.”

八里岗遗址

位于邓州市东约三千米处湍河南岸八里岗西北的坡状高岗上，属于新石器时代的古文化部落遗址，距今约 6800 年，文化层厚三至五米，面积近九万平方米。于 1957 年被发现。

Baligang

Located on the south bank of the Tuanhe River, which is 3,000 meters to the east of Dengzhou City, the Relic of Baligang is the symbol of ancient tribes in the Neolithic age. After a long historical period of 6,800 years, the cultural layer of the tribe relic is as thick as 3 to 5 meters.

楚长城遗址

主要位于南召县和方城县拐河镇，多为土筑，东北至西南走向，东接小擂鼓台山，西连大擂鼓台山，残高 1～1.5 米，宽约四米，全长 500 米。城垣南侧有一古城遗址，后人称之为“霸王城”，村人曾在土城附近发现一枚铜镞，为春秋时期遗物。

The Great Wall of Chu (770BC-221BC)

Mainly located in Nanzhao County and Guaihe Town in Fangcheng County, the northeast-to-southwest running Great Wall of Chu is mostly made of clay. The relic is 1 to 1.5 meters high, 4 meters wide, and 500 meters long.

张衡墓

张衡墓（张衡博物馆）位于南阳市石桥镇小石桥村的西北隅，是我国东汉伟大的科学家、发明家、文学家张衡的长眠之地。

Zhangheng's Tomb (Zhang Heng Museum)

Zhangheng's Tomb is located in the northwest corner of Xiaoshiqiao Village in Shiqiao Town of Nanyang City. Zhangheng was a great scientist, inventor, and scholar during the Eastern Han Dynasty (25-220).

湖阳公主墓

位于唐河县湖阳镇，是著名的历史名胜古迹。湖阳公主刘黄是东汉开国皇帝——汉光武帝刘秀的长姐。

The Tomb of Princess Huyang

The Tomb of Princess Huyang is located in Huyang Town of Tanghe County. Princess Huyang, (Liu Huang) was the elder sister of the first emperor of the Eastern Han Dynasty (25-220), Emperor Guangwu (Liu Xiu) (6BC-57AD).

哪吒庙

相传距西峡县县城 12 千米处的丁河镇奎文村，为《封神演义》中的哪吒太子的出生地。有与哪吒传说有关的陈塘关、杏花村、翠屏山、侍郎村、九湾河等景点。哪吒祖庙背靠山坡，有正殿一座，殿内供奉哪吒太子神像。

Temple of Ne Zha

According to folk stories, Kuiwen Village of Dinghe Town, which is 12 kilometers away from Xixia County, was the birthplace of Prince Ne Zha, a figure in the legend Apotheosis of Heroes. The temple is backed by a mountain. Inside the main hall of the temple stands a statue of Ne Zha for people to worship.

荷花洞

位于西峡县双龙镇东北的伏岭村，独阜岭南麓的峭壁之下，为豫西南较为罕见的石灰岩溶洞奇观。溶洞面积约 3500 平方米，洞高七米。洞顶多处钟乳石形如出水荷花初绽或含苞待放，中间呈红色，周边呈白色，色泽鲜艳，“荷花洞”因此得名。洞内石钟乳林立，形象逼真。

Lotus Cave

Lotus Cave is a karst cave located in Fuling Village, northeast of Shuanglong Town, Xixia County. It has a total area of 3,500 square meters and a height of 7 meters. The cave was named Lotus Cave because several stalactites on the top of the cave are red in the center and white on the rim, resembling lotuses floating on the water.

太白顶

太白顶为桐柏山主峰，海拔 1140 米，又名凌云峰、白云山、胎簪山、大复峰，横跨豫鄂两省，为千里淮河源头。在此登顶远眺，北视中原，南阅楚天，万山俱下，极目千里。

Taibai Summit

Taibai Summit is the peak of Tongbai Mountain with an altitude of 1,140 meters. As the source of the Huaihe River, it spans Henan and Hubei provinces. Here visitors will be able to appreciate a panoramic view of the whole mountain and see as far as they can.

龙潭沟

龙潭沟自然生态风景区位于伏牛山脉腹地的西峡县双龙镇，距西峡县城 30 千米，311 国道途经此地，交通极为便利。景区景点集中，瀑布密集，融山秀、石奇、水澈、林茂、潭幽于一体，被誉为“中原一绝，人间仙境”。

Longtan Valley

Longtan Valley is located in the heartland of the Funiu Mountain Range - Shuanglong Town of Xixia County. This place enjoys easy access to transportation routes. All kinds of beautiful scenery can be found here, including beautiful mountains, rocks of unique shapes, clear waters, lush forest and tranquil pools.

让河

让河隶属于内乡县板场乡，紧挨西去洛阳的 249 省道，距县城约 50 千米。这里山美、水秀、石奇、人和，是南阳市一处度假休闲之地。

Rang River

Located in Banchang Township of Neixiang County, Rang River is 50 kilometers away from the county. This vacation resort in Nanyang boasts beautiful mountains, clear waters and unique stones.

五朵山

位于南召县四棵树乡境内，是南阳伏牛山世界地质公园的主要园区之一，中原地区久负盛名的道教文化圣地。景区总面积 126 平方千米，现已开发休闲戏水暴瀑峡、进香祈福万福宫、登高揽胜五朵峰三大部分。

Wuduo Mountain

Located in Sikeshu Township of Nanzhao County, Wuduo Mountain is a major part of the Funiu Mountain World Geological Park in Nanyang. It has long been reputed as a holy and important cultural site for Taoism.

云华蝙蝠洞

位于西峡县五里桥乡白河村境内，属喀斯特岩溶地貌，因该洞内岩壁上布满云朵样花纹，似锦似缎，堂皇华丽，又因里面有无数蝙蝠栖息，故名“云华蝙蝠洞”。据专家考证，该洞形成于中生代白垩纪，距今有6500万年的历史。

Yunhua Bat Cave

Yunhua Bat Cave is a karst cave located in Behe Village, Wuliqiao Township of Xixia County. Formed about 65,000 years ago, the inner wall of the cave is decorated by colorful cloud patterns and numerous bats live here.

友兰湿地公园

位于唐河西岸，是为纪念哲学大师冯友兰而建，落成时间为2011年11月13日，冯友兰纪念馆位于公园中央。湿地公园北门正对的是唐河县革命纪念馆。

Youlan Wetland Park

Located on the west bank of the Tanghe River, Youlan Wetland Park was completed on November 13th, 2011 in honor of Chinese philosopher, Feng Youlan (1985-1990). A memorial museum for Feng sits at its center.

独山风景区

位于南阳市中心城区东北 3000 米处，跨越面积四平方千米，山体浑圆，海拔 368 米。北连丰紫，西看塔磨，东傍白河，南俯宛城，是南阳的战略要地。独山历史悠久，风景秀丽，是南阳城区群众早春踏青、登山观光、朝山拜佛的传统去处。

Dushan Scenic Spot

With an altitude of 368 meters, Dushan Scenic Spot is located 3,000 meters northeast of downtown of Nanyang. With its long history and beautiful scenery, it is a good place for spring outings and for practicing Buddhist worship.

龙王沟风景区

位于南阳市中心城区正北方向 16 千米左右。总面积为 21 平方千米，其中水域面积 11 平方千米。核心景区为麒麟湖（中心岛），岛上有丰富的森林植被和种类繁多的野生鸟类，空气清新，环境宜人。

Longwanggou Scenic Spot

Located about 16 kilometers to the right, north of Nanyang City, Longwanggou Scenic Spot covers an area of 21 square kilometers, among which 11 square kilometers are covered by water. Zhongxin Island in Qilin Lake is famous for various plants, forests, and wild birds.

淮源风景区

位于南阳市豫鄂交界的桐柏山脉中段，总面积 108 平方千米，是淮河的发源地和江淮两大水系的天然分界线。分为淮源、桃花洞、太白顶、水帘洞四大特色区域，集怀源、盘古、红色、佛道等文化于一身。

Huaiyuan Scenic Spot

Huaiyuan Scenic Spot is in the middle of the Tongbai Mountain Range, spanning the border area of Henan and Hubei Province. As the source of the Huaihe River, it boasts both a magnificent natural and cultural landscape.

紫山风景区

位于南阳市区卧龙区蒲山镇境内，跨越面积25平方千米。有三峰，主峰海拔为326米。山体构造为黑云母花岗岩，沟壑幽深，怪石林立，有“紫灵耸翠”美誉。紫山物华秀丽，名胜古迹颇多，如凤雏台遗址，为汉末庞统隐居之地。佛教氛围浓厚，顶端有祖师庙。

Purple Mountain Scenic Spot

Located in Pushan Town, Wolong District of Nanyang, Purple Mountain consists of three peaks, with the major peak having an altitude of 326 meters. Cliffs and unique rocks can be found in the beautiful scenery.

五道幢

位于西峡县城东北部二郎坪乡境内，是南阳伏牛山世界地质公园的核心地带。五道幢景区峡谷奇特，谷中河水清澈，山水相映相辉，如诗如画。高瀑、幽潭、翠山、密林、浅滩、奇峰形成了这里独特的自然美景。

Wudaochuang

Located in Erlangping Township in the northeast area of Xixia County, Wudaochuang is in the center of the Funiu Mountain World Geological Park in Nanyang. There are various attractions here, including high waterfalls, tranquil pools, green hills, dense forests, shoals and uniqely-shaped mountain peaks.

桐柏革命纪念馆

位于桐柏县城南叶家大庄。由前国家主席李先念亲笔题写馆名，整体占地两万多平方米。纪念馆所在地原为当地名绅叶逢雨先生在清嘉庆年间（1796 ~ 1820 年）的住宅，叶家在三军会师桐柏后，主动腾出房间作为党机关的办公场所。馆内展列了革命战争时期柏桐县有关的资料、照片、实物等。

Tongbai Revolution Memorial Museum

Located in Ye's Village, south of the Tongbai County seat, the Tongbai Revolution Memorial Museum was once the residence of Ye Fengyu, a local squire during the reign of Emperor Jiaqing (1796-1820) in the Qing Dynasty (1644-1911). There are many materials, pictures, and objects related to the revolutionary years on display here.

编外雷锋团

1989 年，与雷锋一起入伍的 560 名邓州籍复员回乡战士自发成立了“学雷锋指导委员会”，之后邓州先后成立学雷锋指导小组 23 个，学雷锋送温暖小分队 1300 多个，为人们架起了一座了解雷锋的桥梁，群众亲切地称他们为“编外雷锋团”。1997 年正式成立邓州“编外雷锋团”。

Informal Lei Feng Group

Lei Feng (1940-1962) was a communist solider well-known in China for his benevolence and readiness to help others. In 1989, 560 soldiers from Dengzhou, who were enlisted at with Lei Feng, established a local volunteer group to carry on Lei Feng's spirit after he died.

南阳庙会

南阳的庙会，有着长久的渊源。根据出土的汉画像石及文献资料来看，它至少在春秋战国时期即已出现。庙会多在春、秋季举行，由求寿、拜佛、进香、还愿兴起。南阳庙会是重要的敬神、娱乐和文化交流形式。著名的有农历三月三独山庙会等。

Nanyang Temple Fair

According to research, the temple fair has a history of over 2,000 years. Normally it is held in spring or autumn in order to pray for longevity, practice Buddhism, offer incense or redeem a vow. The Dushan Temple Fair is on March 3rd of the lunar calendar and is a significant event here.

南阳烙画

烙画亦称烫画、火笔画，起源于西汉，是“南阳三大宝”之一。它是用温度在 300～800 摄氏度的铁扦代笔，利用碳化原理，在竹木、丝娟、宣纸等材料上作画，把绘画艺术的各种表现技法与烙画艺术融为一体，有较高的艺术价值。

Nanyang Pyrography

With a history of over 2,000 years, pyrography is a drawing technique that uses iron needles that are heated to a temperature of 300-800 degrees centigrade. Pyrographic drawings can be made on many materials, such as bamboo, silk, and xuan paper (a high quality paper made of rice).

镇平玉雕工艺

据考证，镇平玉雕生产起源于西汉，宋、元渐具规模，明清以来成为本县的一大产业。以独山玉和岫玉为主要原料，工艺精湛，造型逼真。自 1993 年起连续举办八届“中国镇平国际玉雕节”，名声大震。此图为玉雕之城——镇平。

Jade Carving in Zhenping

Jade carving here boasts a history of over 2,000 years. With Dushan jade and Xiu jade as their main materials, the craftsmen apply sophisticated technology to create lifelike, artistic works.

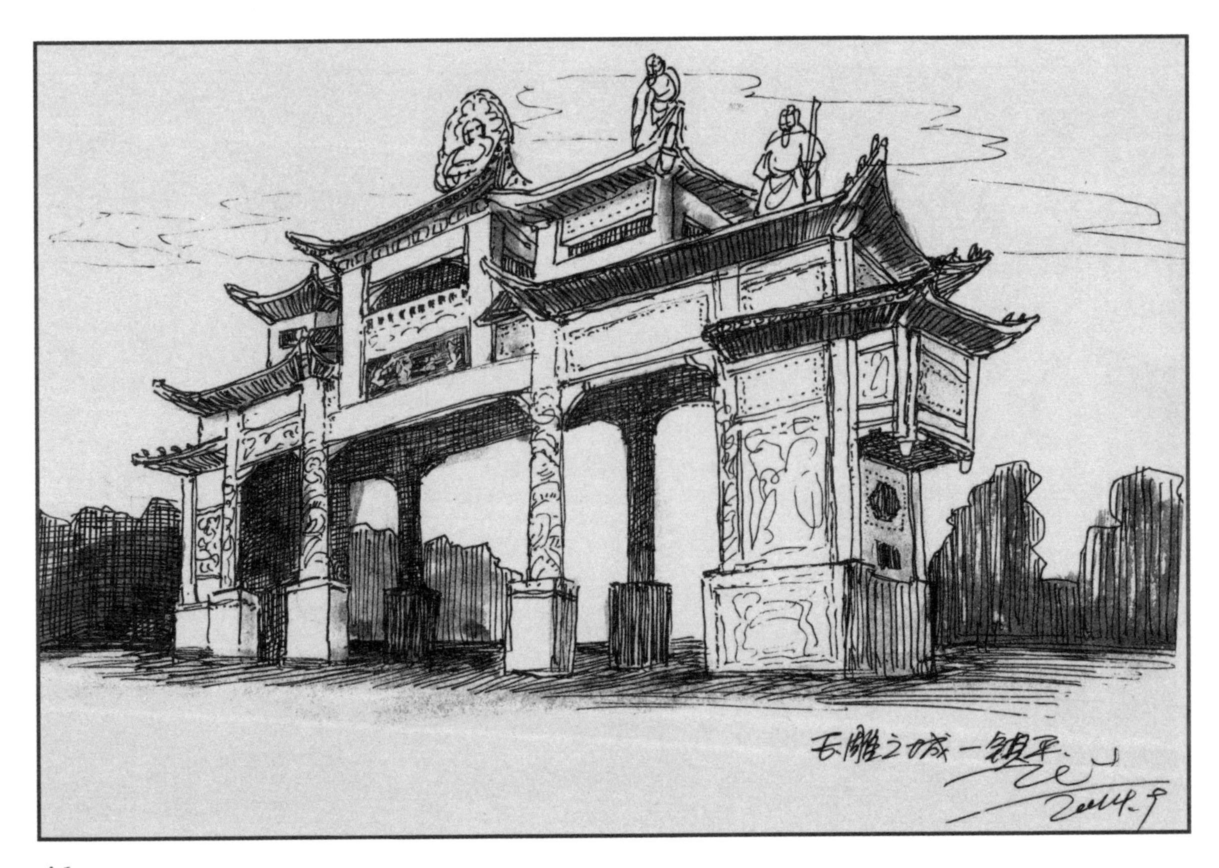

丹江号子

淅川民歌的一种。是淅川船工在千百年来拉纤行船的劳动中，集体创作出来的民间歌曲。最初产生于船工们劳动时统一喊号，是船工们在拉纤、撑篙、推舟、扛船劳动中，一人领、众人和的一种曲调形式，不需要任何乐器伴奏。号子音韵悠长，雄浑有力。

Danjiang Work Song

The "Danjiang work song" is a style of folk song in Xichuan. It originates from songs once sung by boatmen when towing a boat to the bank and usually has one lead singer. This kind of song is sung a capella.

南阳柞蚕丝绸

南阳历来有“养蚕为业，植柞为本”的传统。据记载，南阳柞绸始于汉代，所产丝绸质地精良、图案美观、灿若云锦。清末，南阳柞绸业进入鼎盛期，远销美英等国。清光绪二十七年（1901 年），上海“久成”丝绸行在李青店、板山坪设立分行。

Nanyang Tussah Silk

It was recorded that Nanyang Tussah silk originates from the Han Dynasty (202 BC-220) and reached its prime in the Qing Dynasty (1644-1911). With a history of over 2,000 years, it has long been favored for its good quality and exquisite patterns.

南阳三弦书

说唱艺术，又称铰子书、腿板书，已有250多年历史，因用三弦、铰子（小铜钹）为主要伴奏乐器而得名。初为一人怀抱三弦，腿束脚板自弹自唱。后演变为演唱者手击铰子或八角鼓，另有专门抱三弦和坠胡的进行伴奏并在演唱中帮腔、插话而成为二三人一台戏。

Sanxianshu (storytelling accompanied by three-string plucked instrument) in Nanyang

This is an old form of storytelling with a history of more than 250 years. The storyteller tells stories with songs while playing two pieces of metal or an octagonal drum, accompanied by three-strings and two-string plucked musical instruments.

方城石猴

源于方城县独树镇砚山铺村的一种传统石雕艺术品。民间艺人采用当地石料，雕出猴子形状，然后用黄、绿、红、黑等颜料涂染在石猴身上而成，故又称“画石猴”。起源于宋，鼎盛于明清。因与“时候”谐音，意指“好运气”，逢年过节便以此为吉祥物给孩子佩戴。图画为民间艺人正在雕刻石猴。

Stone Monkey of Fangcheng

First created by people from Yanshanpu Village, Dushu Town of Fangcheng County, this carving technique is also called “Painting on Stone Monkey.” Folk artists carve out the shape of a monkey on a local stone and then paint it with different pigments. It dates back to over 1,000 years and is seen as an auspicious gift given during festivals.

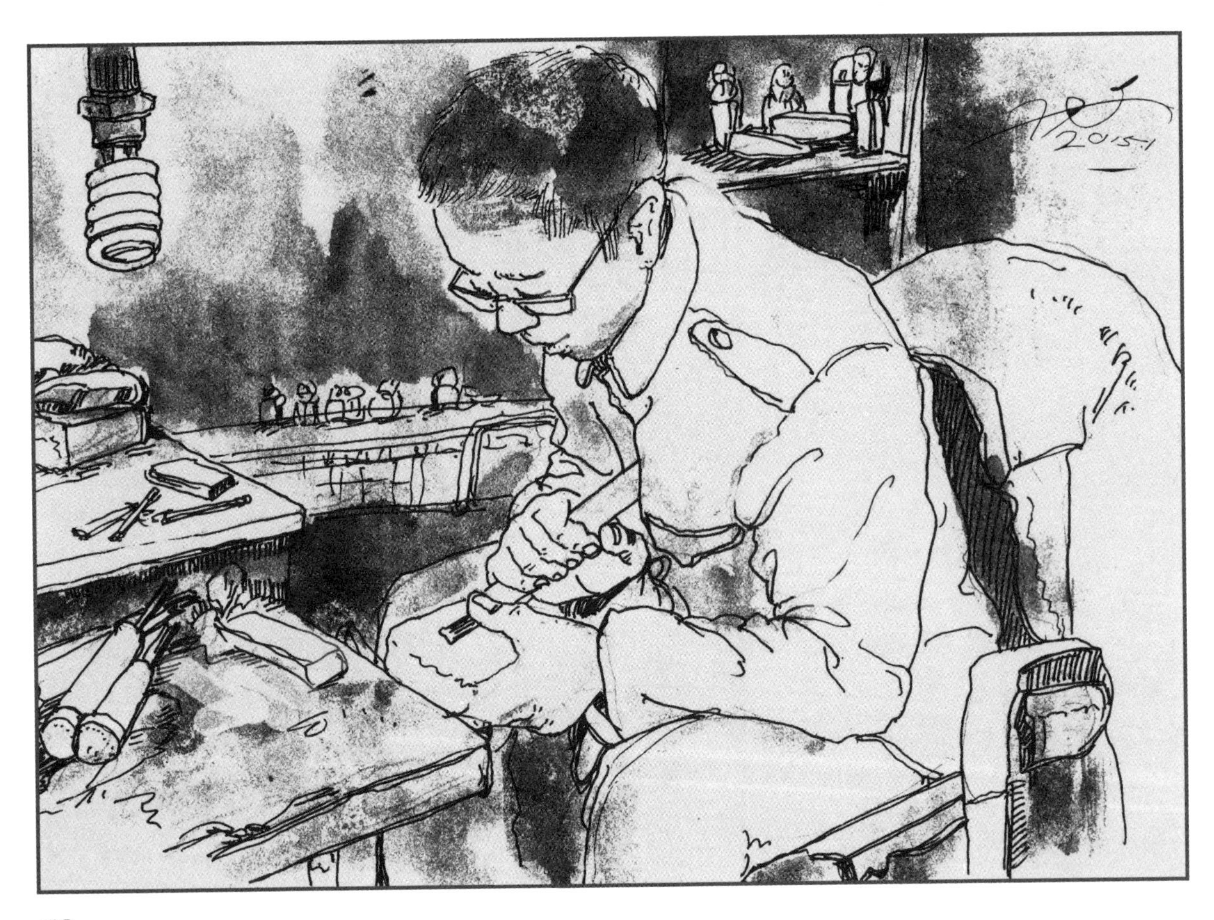

新野猴戏

俗称耍猴儿，已有 2000 年的历史。明清时期，新野民间玩猴就已经较为流行。近年新野出土的大量汉画砖上，除了杂技、游戏之外，猴子、狗和人在一起狩猎、嬉戏的精彩画面屡见不鲜。

Monkey Show in Xinye

Known as Shuahou'er among the locals, the monkey show has a history of over 2,000 years. It became quite popular in Xinye during the Ming (1368-1644) and Qing (1644-1911) dynasties. On many of the painted bricks from the Han Dynasty (202BC-220) discovered in recent years, scenes of hunting are depicted in which monkeys and dogs are often seen hunting and playing with the hunters.

送财神

南阳地区春节时期的古老民俗文化。一种是手拿着一张纸印的财神在门外嚷着："送财神爷的来啦！"这时屋里的主人，便拿赏钱给来人以示欢迎财神。或者是装扮成财神爷的模样，身穿红袍，头戴纱帽，嘴上挂着假胡子，身上背着一个收钱的黄布袋，后面跟着几个敲锣打鼓的，挨家挨户地去散发财神爷像，以便讨赏钱。

Sending Portraits of God of Wealth

It is an ancient folk practice during Spring Festival in Nanyang to send portraits of the God of Wealth to others. Some local people disguised as the God of Wealth will send portraits of the god from door to door and ask for a monetary reward.

南阳民歌

多反映劳动、生活、爱情和劳动者对旧社会的反抗等内容，有着浓厚纯朴的生活气息；体现着南阳地区人民的精神世界，表达了他们的思想感情和愿望。曲调优美动听，唱腔圆滑流畅、灵活多样，且易记易学易传。

Folk Song of Nanyang

The topics of these folk songs are mostly related to work, life, and love. In these songs, a strong sense of rustic life can be felt. The singing is melodic and smooth and the song styles are diverse.

西峡重阳节登高

每年农历九月九重阳节，西峡及周边民众到重阳寺，颂扬李娘娘和重阳公主，祭案上摆重阳糕、菊花酒、茱萸枝，然后扶老携幼至菊花山、佛爷山、云彩山等登高望远；祭典结束后，家家都要在当日大吃重阳糕、痛饮菊花酒。据记载，隋唐时期，因这里重阳节俗盛，菊花山名声远播，曾专设“菊潭县”。

Mountain Climbing in Xixia During the Chung Yeung Festival

People visit the Chung Yeung Temple during the Chung Yeung Festival on September 9th of the lunar calendar every year. They do climbing, yelling (which the locals believe to be good for health), drink Chinese chrysanthemum wine, eat Chung Yeung pastries, and wear Zhuyu sachets (made from an herb usually used by the Chinese for expelling worms) as part of their celebration. The custom has been passed down from generation to generation for thousands of years.

南阳大鼓

又称鼓儿哼、鼓儿词、南阳鼓词、毂辘词等，是南阳市戏曲剧种之一。源于唐代的道调、道曲。20 世纪 20 年代是鼓儿词的兴盛时期，以镇平县为集中地，主要分布在南阳、社旗、唐河、新野、方城等地。后来在表演中加入三弦伴奏，并改进唱腔，称为“南阳大鼓”。

Nanyang Drum

Also known as Guerheng, Nanyang Drum is one of the local operas in Nanyang City. Originated from Daodiao or Daoqu in the Tang Dynasty (618-907), three-string plucked instruments were later added to the accompanying music and its singing styles were also improved.

驻马店市

Zhumadian City

南海禅寺

位于汝南县城东南隅，始建于明嘉靖二十四年（1545 年），占地约 30 万平方米。寺域开阔，建筑浩繁。主体建筑大雄宝殿，殿基两米，平面呈边长 80 米的正方形，建筑面积达 6400 平方米，号称“亚洲第一殿”。

Nanhai Buddhist Temple

Located in the southeast corner of Ru’nan County, Nanhai Buddhist Temple was first built in 1545 with an area of about 300 thousand square meters. The main part, Mahavira Hall, covers an area of 6,400 square meters.

北泉寺

位于驻马店市西北的乐山、秀山之间，始建于北齐年间，已有1600多年历史，最初叫天宫。因确山县自南而北有南泉、中泉、北泉三泉，泉前皆有寺院，该寺院位居北泉，故又称北泉寺。坐北向南，院内有大佛殿、二佛殿等建筑。

North Spring Temple

Sitting between Leshan Mountain and Xiushan Mountain in the northwest of Zhumadian City, the temple boasts a history of over 1,600 years. From south to north, there are three famous springs in Queshan County: South Spring, Middle Spring, and North Spring. The temple got its name because it is located near North Spring.

悟颖塔

位于汝南县城南，因唐代和尚悟颖所建而名悟颖塔。又有传说每年夏至日中午没有影子，称无影塔。塔高26米，整个塔体外廊呈抛物线形。塔身基座上刻有“隆庆元年崇藩施财重建宝塔”字样。塔身古朴美观，建造结构合理。

Wu Ying Tower

Wu Ying Tower was built in the south of Runan County by monk Wu Ying during the Tang Dynasty (618-907). It is said that it leaves no shadow on the ground at noon on summer solstice every year, so it is also called "Shadow-Free Tower." It is 26 meters high and has a parabolic outline.

新蔡文庙

始建于元大德八年（1304 年），自李演创建于街北，至明嘉靖三十三年（1554 年）朱令如复迁于古城东南隅，两朝二百余年三易其地。文庙内有大成殿一座，殿内有孔子铜像一尊，文庙旁设儒学，建明伦堂一座，是讲学之所。

Xincai Confucious' Temple

Xincai Confucious' Temple was built in 1304 and was once used as a place to disseminate Confucianism. In the Dacheng Hall, there is a bronze statue of Confucius.

伏羲画卦亭

位于上蔡县城关东蔡冈（现白圭庙一带），从清康熙年间（1662 ~ 1722 年）尚存的蔡邕题的碑刻“伏羲蓍台”四字，推测其为东汉之前的建筑。该八角凉亭重檐高翘、筒瓦盖顶、古雕秀丽，矗立于约两米高的台基之上，掩映在几株古柏之中。后又建一座伏羲庙。

Pavilion where Fu Xi Painted the Eight Diagrams

This pavilion is not far from Baigui Temple of Shangcai County and it is supposed to have a history of over 2,000 years. It is an octagonal pavilion decorated with beautiful carvings and eaves pointing to the sky. (Fu Xi was the ancestor of mankind according to China's classic myths.)

蔡明园

位于上蔡县蔡都镇西南部，占地约70多万平方米，综合性公园。正门上方南面写有“蔡明园”、北面写有“回头是岸”。大门石块雕有四大天王、八大金刚、飞天、花、鱼、伞等精美图案。

Caiming Park

Caiming Park is a comprehensive park located in the southwest of Caidu Town of Shangcai County. Stones on its doors are decorated with delicate carvings of flowers and fish.

大吕书院

位于新蔡县古吕镇老城东北隅，现为“今是中学”所在地。这里原为甘泉寺，因后院有眼涌泉，其水清洌甘甜，时人建庙祭之。明嘉靖年间（1522 ~ 1566 年），王紘筑台凿洞，栽花植松。其子王惟善又垒石为山，增建亭榭，命名曰“望岳园”。有不少文人骚客在此聚会，咏诗作赋。明末被毁。

Dalü Academy

Located in the northeast corner of old Gulü Town in Xincai County, Dalü Academy was originally named Ganquan Temple because there used to be a spring in the backyard, and people in the town built a temple to worship it. During the Ming Dynasty (1368-1644), Wangyue Park was built here and later becamethe preferred venue for intellectuals and poets to share their literary works. Unfortunately, the park was destroyed at the end of the dynasty.

弘济桥

位于汝南县城北，跨汝河。初为木桥，后被洪水冲走。明弘治十八年（1505 年），改建为石桥，明万历十年（1582 年）又重修。桥面长 44.6 米，宽 6.5 米，大券跨度为 24.8 米。在大券两边附两个小券，美丽大方，雄伟壮观。

Hongji Bridge

Located in the north of Runan County, the wooden Hongji Bridge was rebuilt into a stone bridge in 1505. It is 44.6 meters long and 6.5 meters wide, with a large bridge arch spanning 25 meters. The big arch is accompanied by two smaller ones, making the whole structure quite magnificent.

月旦亭

位于平舆县城小清河岛上，为东汉遗址。东汉末年名士许劭、许靖兄弟在家乡讲学，评论时事人物。每月第一天举行，故称“月旦评”，岛上的草亭被呼作“月旦亭”，“月旦人物”也成为品评人物的一个成语。遗址面积近一万平方米。

Yuedan Pavilion

Located on Xiaoqinghe Island of Pingyu County, Yuedan Pavilion boasts a history of over 2,000 years. During the late Eastern Han Dynasty (25-220), brothers Xu Shao and Xu Jing began to give lectures in their hometown to discuss social issues. The lectures were given on the first day of every month, therefore it got the name Yuedan Commentary ("Yuedan" means the "beginning of the month"), and the pavilion on the island was called Yuedan Pavilion.

天中山碑

位于汝南县城，唐朝大书法家颜真卿被困蔡州时书写了“天中山”三个大字，后来颜真卿被叛将李希烈所杀。后世为纪念他舍生取义的精神，在汝南城内修建颜鲁公祠，将他生前亲书的“天中山”镌刻立碑。今已建成汝南县天中山公园。

Tianzhongshan Stele

The Tianzhongshan Stele is located in Runan County. Yan Zhenqing, a great calligrapher of the Tang Dynasty (618-907), wrote three characters, “tian zhong shan,” on the stele while he was surrounded by his enemy in Caizhou. He was later killed by traitors. In honor of Yan, an ancestral hall was built in Runan, and the characters written by the great calligrapher were engraved on a stele.

宝岩寺塔

位于西平县城东关，原在宝岩寺外西侧，故又称“宝严寺塔”“东关塔”。始建于北宋时期，仿木结构楼阁式七级砖塔，高 28.8 米；平面呈六角形；塔角雕饰龙首，塔身浮雕壁画；顶端有铁铸塔刹，呈莲花状，高两米。

Pagoda of Baoyan Temple

Located at the east entrance of Xiping County, the pagoda has a history of over a thousand years. It is a seven-floor brick pagoda, with a height of 28.8 meters. This is a hexagonal building with a two-meter, lotus-shaped spire on the top.

后龙亭

位于汝南县龙亭街，始建时间无考，清代重修。建筑在两米高的砖砌台基上，硬山灰筒瓦顶，长 11.3 米，宽十米，高三米，原是金哀宗行宫旧址。后在战争中遭到破坏，1986 年于旧址上重建，为钢筋水泥结构，面积 200 平方米。

Houlong Pavilion

Houlong Pavilion is located on Longting Street of Runan County. It was rebuilt in the Qing Dynasty (1644-1911) and the date of its original construction is not clear. Built on a two-meter brick pedestal, it is 11.3 meters long, 10 meters wide, and 3 meters high.

普照寺塔

又称秀公戒师和尚塔，位于平舆县李屯乡柳屯村。寺为金皇统年间（1141 ~ 1149 年）所建，天德年间（1149 ~ 1151 年），住持秀公戒师和尚圆寂后，弟子广全等于金明昌五年（1194 年）为其建造。如今寺毁仅存塔。塔通高 14 米，砖雕细致精美。具有极高的建筑审美和考古价值。

Pagoda of the Puzhao Temple

The pagoda is also known as the Pagoda of Monk Xiugong. Located in Liutun Village, Litun Township of Pingyu County, it was first built in 1194. With a height of 14 meters, it is an exquisite building featuring brick carvings.

蟾虎寺

位于蔡国故城西卧龙岗上，始建于东汉明帝时期，是我国最早的寺院之一，占地约 1.5 万平方米。这里既是新石器时代中原仰韶文化、龙山文化和湖北屈家岭文化的遗址，又是周公旦东征伐蔡的古战场和蔡侯叔度被擒后的软禁地。从蟾虎寺可远观古县城。

Chanhu Temple

Located on the Wolong Hill, west of the capital city of the Cai Kingdom, the temple boasts a history of 2,000 years. It is an important relic of Neolithic central China (9000 BC-2000 BC) and a famous ancient battlefield. Visitors can have a bird's eye view of the ancient county from the temple.

汉安成故城遗址

位于正阳县寒冻镇固城寺村，是距今有2000多年的历史名城，始建于西汉高帝六年（前201年），遗址内有高大方形的土岭、大而深的河沟仍能看出当时安成的繁荣。

Ancheng Town in the Han Dynasty

Located in Guchengsi Village, Handong Town of Zhengyang County, this relic has a history of more than 2,000 years. Huge mounds and ditches leave traces of the prosperity in Ancheng from centuries ago.

古江国都城遗址

位于正阳境内，淮河流域一带，其都城遗址在今大林乡涂店村，东西长两千米，南北宽 1.2 千米。在古城遗址内，曾出土有商、周时代生活用陶器残片。周襄王二十九年（前 623 年），楚国灭掉江国，江国子孙出逃流落外地，以国为姓，始有江姓。

The Capital City of the Ancient Jiang Kingdom

The capital city is located in Tudian Village, Dalin Township of Zhengyang County. It is 2 kilometers long from east to west and 1.2 kilometers wide from north to south. In 623 BC, when the ancient Jiang Kingdom was defeated by the Chu Kingdom, the offspring of Jiang fled to other areas, and then began to use the name of their former kingdom as their family name.

付寨闰楼商周古墓葬群

位于正阳县付寨村，商周时期的贵族及平民墓群，主要分布在文殊河的南岸冈地上。经考古发现，古墓群共有商周直至明清不同时期的墓穴 149 座，总发掘面积 1.8 万平方米，出土文物证明此地带曾是一处高度文明的商代部落遗址。

Runlou Ancient Tombs of the Shang Dynasty (1600BC-1046BC) and Zhou Dynasty (1046BC-256BC) in Fuzhai

The 149 tombs dating back to various historical periods, which were discovered by archaeologists in Fuzhai Village, Fuzhai Township of Zhengyang County. The relics prove that the surrounding areas used to belong to an ancient and cultivated tribe of the Shang Dynasty (1600 BC-1046 BC).

新蔡故城

新蔡县位于河南省东南部。夏朝初年，伯夷辅佐大禹治水有功被封为吕侯，在此地建立吕国。周景王十六年（前 529 年），蔡平侯将国都迁至此，史称新蔡国。故城东西呈长方形，占地 580 余万平方米，故城内文化层较厚，文化遗存丰富。

Xincai Kingdom

Originally known as the Ancient Lü Kingdom, Xincai County is located in the southeast of Henan province. Prince Caiping moved the capital of the Ancient Lü Kingdom to Xincai around 529 BC and named the kingdom Xincai. The rectangular site extends from the east to the west, boasting numerous cultural artifacts.

干宝故里

干宝，字令升，新蔡县佛阁寺镇小干庄人。东晋著名的文学家、史学家。东晋初年，经王导推荐，领修国史。历任山阴令，始安太守司徒右长史，散骑常侍等职。堪称中国小说之鼻祖，著作有《晋记》《搜神记》等。

Hometown of Gan Bao

Gan Bao was a resident of Xiaoganzhuang Village, Fogesi Town of Xincai County. He was a famous historian and scholar during the Eastern Jin Dynasty (317-420). As the forerunner of Chinese novels, he wrote masterpieces like *History of Jin* and Searching for Deities.

梁祝墓

位于汝南县马庄乡古官道两侧。梁山伯与祝英台传说为民间传说，被称为东方的“罗密欧与朱丽叶”。因梁山伯与祝英台并未订婚而分葬墓，符合当时的风俗习惯。汝南县被称为“中国梁祝之乡”，马乡镇更名为梁祝镇。

Tombs of Liang Shanbo and Zhu Yingtai

The two tombs were located on two sides of the ancient road in Mazhuang township of Runan County. Known as the Romeo and Juliet of the East, Liang Shanbo and Zhu Yingtai are household names. As they were not engaged when they died, they were buried in separate tombs. Runan County is known as the birthplace of this love story.

战国冶铁遗址

位于西平县酒店乡酒店村，是战国至汉时期重要的冶铁基地。遗址呈长方形，东西长 558 米，南北宽 190 米，文化层厚 1.5 米。遗址上残留有炼炉残壁和生活用陶盆罐及砖瓦材料等。另在遗址南尚存冶炼炉一座。

Iron-Semlting Site of the Warring States Period (475BC-221BC)

Located in Jiudian Village, Jiudian Township of Xiping County, this was a site for iron-smelting during the Warring States Period (475 BC-221 BC) and the Han Dynasty (202BC-220). It is 558 meters long and 190 meters wide. A smelter is preserved in the southern part of the old site.

子路问津处

位于新蔡县城南津关渡口。子路是孔子的得意门生，周敬王三十年（前 490 年），孔子与学生云游至此，因汝河挡道，只好让子路向田里干活的两个老农问路。这两个老农竟是长沮和桀溺，都是当时有名的隐士。孔子由此感叹此地多睿智隐士。

Ferry where Zi Lu Asked Directions

The ferry, called Nanjinguan, is located in Xincai County. Zi Lu (542 BC-480 BC) was one of the best students of Confucius. When Confucius and his students traveled to the ferry, Zi Lu was sent to ask the way from two peasants on a farm, who were discovered to be prominent hermits later on. Therefore, this area was praised by Confucius as a place of wise recluses.

重阳登高处

位于上蔡县城北，东汉上蔡人桓景为避祸消灾而于九月九日举家佩茱萸绛囊，登高于望河楼饮菊花酒。由此兴起，盛于唐宋，经过近 2000 年的演变、发展，现已成为普天之下尊老敬老的华人之节。上蔡县被称为“中国重阳文化之乡”。

Mountain Climbing Site during Chung Yeung Festival

This site is located in western Shangcai County. Huan Jing, a resident of Shangcai in the Eastern Han Dynasty (25-220), along with his family, climbed to Wanghelou and had chrysanthemum wine there in hopes of avoiding evil spirits and misfortunes and pray for the longevity of his parents. After over 2,000 years, the Chung Yeung Festival, which falls on the ninth day of the ninth month of the lunar calendar, has become a day to pay reverence to and pray for senior citizens.

雷岗战役遗址

1947年8月24日，刘邓大军千里跃进大别山，途径正值汛期的汝河，在紧急情况下，司令员果断提出了“狭路相逢勇者胜”的战斗口号，激励全军斗志，突破汝河雷岗防线，取得了雷岗战役的重大胜利，顺利挺进了大别山。中国革命由此发生了巨大转折，也使正阳雷岗闻名遐迩。

Leigang Battle

On the march toward Dabie Mountain, the Liu Deng army went through the Ruhe River and defeated the enemy in Leigang on August 24, 1947.

宿鸭湖

位于汝南县罗店乡东，北起玉皇庙，南至野猪岗，东临桂庄，西到别桥。始建于 1958 年，是我国人工修建的面积最大、堤坝最长的平原水库。以防洪为主，结合灌溉、发电、养殖、旅游等多项目综合利用的大型水利枢纽工程。

Suya Lake

The man-made lake is located to the east of Luodian Township of Ru'nan County. It was first built in 1958 and is a reservoir on plain with a huge water area and a long dam.

嵖岈山风景区

位于遂平县境内，系伏牛山东缘余脉，又名玲珑山、石猴仙山。山势嵯峨，怪石林立，湖山相映。景区人文史迹星罗棋布，枚不胜举。有九大奇观、九大名峰、九大名洞、九大名棚、九大奇石，集“奇、险、奥、幽”于一体，享有“华夏盆景”“伏牛奇观”之美誉。

Chaya Mountain Scenic Spot

Also known as Linglong Mountain or Shihouxian Mountain, Chaya Mountain is an offshoot of the eastern range of the Funiushan Mountains in Suiping County. Surrounded by clear lakes and steep mountains, the spot is full of fantastic rocks and peaks.

龙天沟风景区

位于遂平县境内。南与嵖岈山风景区比肩而立一脉相连。面积约两万平方米，是伏牛山余脉，有中原“九寨沟”之美誉。景区内幽深恬静，动植物资源丰富，原始植被保存完好，古遗址遗迹较多。

Longtiangou Scenic Spot

This scenic spot is an offshoot of Funiushan Mountain in Suiping County. It is famous for its serene environment, various fauna and flora, well-preserved primeval vegetation, and large numbers of cultural relics.

棠溪源风景区

位于西平县境内，总面积 38 平方千米，因柏皇始祖（祖之源）、冶铁铸剑（剑之源）、棠溪源头（水之源）而闻名。森林覆盖率 95% 以上，有棠溪湖、棠溪峡、蜘蛛山、跑马岭四个游览区，集绮丽秀美的自然风光与厚重深沉的炎黄文化于一体。

Tangxiyuan Scenic Spot

Located in Xiping County, this scenic spot covers an area of 38 square kilometers. With a vegetation coverage rate of over 95 percent, it boasts magnificent natural scenery.

奚仲公园

位于平舆县城西部，北邻驻新路，西、南临小清河。占地20多万平方米，其中月牙湖一万多平方米，如意湖三万平方米。奚仲是公元前两千多年的古薛国的发明家、政治家，曾发明造车，人们为纪念他而建。

Xizhong Park

The park is located in the west of Pingyu County. Xi Zhong was an inventor and politician 2,000 years ago. He once tried to invent carts.

太任公园

位于平舆县城西南部。太任是有史记载的平舆地域文化的早期代表人物之一，也是我国历史上有贤德之名的母亲之一。她是殷商时期挚国（今平舆）国君任侯的二女儿，从小就培养了高尚的品德，芳名远播。后嫁周先王季历为妻，爱护百姓，尊礼重德。

Tairen Park

Tairen Park is located in the southwest of Pingyu County. Tai Ren is a historical figure of the Zhou Dynasty (1046BC-256BC) who was famous for her moral qualities. She married Prince Zhouxian, Ji Li. She showed great love to the common people and attached great importance to morality.

老乐山

位于确山县城西北，由高低不同的九座山峰组成，山北崖有老虎洞。前有陡峭的十八盘，山顶有清澈甘甜泉水一池，因形状似蛙，故名“蛤蟆泉”。山上还有宏伟的道家真武庙、宣坛庙、拜台宫、玄都宫。每年春季在此举行庙会，朝拜者数以万计。

Laole Mountain

Laole Mountain is located to the northwest of Queshan County and consists of nine hills of various heights. On the top of the mountain, there is a spring and several Taoist temples. Temple fairs are held here every spring.

竹沟革命纪念馆

位于确山县竹沟镇延安街，始建于 1956 年，由周恩来总理题写馆名。是全国建立较早的革命纪念馆之一。馆内有革命旧址 31 处，文物、文献、图片等近千件。是全国 " 红色旅游 " 经典景区之一。

Zhugou Revolution Memorial Museum

Located on Yan'an Street in Zhugou Town of Queshan County, the museum was built in 1956. The words on the plaque above the front door were written by Premier Zhou Enlai. As one of the early revolution memorial museums, it has 31 revolutionary sites and about a thousand artifacts.

杨靖宇将军纪念馆

位于驻马店市驿城区古城乡李湾村，始建于 1966 年秋，建筑面积为 4466 平方米，此地是杨靖宇的故居。

Memorial Museum for General Yang Jingyu

Located in Liwan Village, Gucheng Township, Yicheng District of Zhumadian City, this museum was the residence of General Yang Jingyu (1905-1940), a national hero during the Anti-Japanese War (1937-1945).It was built in the autumn of 1966.

平舆高跷

又叫“踩高跷”，多在春节、正月十五表演。多由男性扮为女装，身穿戏服脚踩高跷，进行表演。据记载高跷早在先秦已在民间流行，已有千年历史。乡土气息浓厚，形式奇特别致，场面热烈，深受民众喜爱。

Stilts in Pingyu

Stilt-walking in Pingyu has a history of over a thousand years, and stilt performers are usually seen during Spring Festival and the Lantern Festival (the fifteenth day of the first lunar month). Most players are men disguised as women, wearing female clothes and makeup.

汝南天中麦草画

是流行于汝南县罗店乡的民间工艺美术品，风格多样，栩栩如生。以板材作基础，用麦草作原料粘贴画稿，并通过熏、蒸、烫、漂等十几道艺术加工处理而成。在风格上有工艺画、变形画、写意画，被授予“中国民间艺术一绝”。

Tianzhong Straw Painting in Runan County

Straw painting is a folk art widely spread throughout Luodian Township of Runan County. Artists paste straws on paintings already drawn on a board, and they have to go through dozens of procedures before the painting is finished (including fumigating, steaming, scalding, rinsing, etc).

西平铜器会

西平民间大铜器是一种独具风格的闹年器乐。历史悠久，流传广泛，距今已有1400多年的历史。素有“城东喇叭、城西铜器”之说，全县80%的自然村都有铜乐队，是群众参与广泛的一种民间艺术表演形式，也是非常值得保护和传承的优秀民间文化。气势恢宏，韵律铿锵。

Brass Instrument Show in Xiping County

Playing brass instruments is a unique tradition of the local people in Xiping to celebrate Spring Festival. It dates back to 1,400 years ago and over 80 percent of villages in Xiping have their own bands playing these instruments.

确山打铁花

是一种民俗娱乐活动，起源于北宋。在空地搭一座三米高的“花棚”，棚顶铺一层绑有烟花、鞭炮的柳树枝。棚顶正中竖一根绑有烟花、鞭炮的杆子，棚旁立一口熔铁的熔炉。打铁花者在棚下手持盛有铁汁的和未盛铁汁的“花棒”互相敲击，铁汁遇到棚顶的柳枝迸散开来，又点燃了棚上的烟花和鞭炮，铁花飞溅，鞭炮齐鸣，十分壮观。

Striking Iron Flowers in Queshan Mountain

This local style of entertainment has a history of over thousands of years. Willow branches were knitted into a net three meters above the ground in an open space. On the net fireworks and firecrackers were tied. Melted iron was pour into hollow, short thick willow branches. The players rushing into under the net and each stroked an iron-filled stick and a normal stick, tossing out the iron which ignited the fireworks and firecrackers. The spilled iron and the sparks from the exploded fireworks and firecrackers were very beautiful.

信阳市
Xinyang City

陈将军祠

位于固始县陈集乡，为“开漳圣王”陈元光祖祠。建于唐天宝年间（742～756 年），现存为清代建筑。坐北朝南，占地 800 平方米。为硬山式砖木结构，石基、石柱、青转灰瓦、堂室含廊。建筑布局严谨，注重细节，石柱下有石狮，做工细致，形象精美，具有较强的地方色彩。

Ancestral Temple of General Chen

The Ancestral Temple of General Chen belonged to Chen Yuan'guang in Chenji Township of Gushi County. It was built during the Tianbao Period (742-756) of the Tang Dynasty (618-907). The existing structure dates back to the Qing Dynasty (1644-1911). The south-facing temple is a traditional building with a stone base, and it is decorated with black bricks, gray tiles, and stone lion statues - through which a strong local sense can be felt.

妙高禅寺

位于固始县西九华山半山腰，始建于隋，曾被称为大佛寺、华岩寺、地藏寺。自明代释祖春卓锡该寺后，成为临济宗大悟山派之祖庭，自成立以来，有文稽考者，已兴传了56代。明末清初，著名高僧释竺启和尚重修妙高寺，使其成为附近最负盛名的佛教圣地。

Miaogao Buddhist Temple

Located halfway up Jiuhua Mountain to the west of Gushi County, the temple has a history of over a thousand years. In the late Ming (1368-1644) and early Qing dynasties (1644-1911), it was revamped by Shizhuqi, an eminent monk at that time.

云霄庙

又称大山奶奶庙，位于固始县安阳山浮光顶峰，为“开漳圣王”陈元光之孙陈酆于唐天宝年间（742～756年）为纪念陈元光祖母魏敬而建。魏氏曾协助平叛盗匪、救助贫民、挂帅出征，立下功绩。

Yunxiao Temple

Located on Fuguang Peak of Anyang Mountain in Gushi County, the temple is known as "Granny Temple in Mountains." It was constructed in honor of Wei Jing, the grandmother of General Chen Yuanguang, who has the legacy of being a virtuous and beloved woman among the locals.

大王庙

又称金龙寺。位于罗山县竹竿镇河口村北街，地处竹竿淮河入河口，始建于元末明初，距今有 800 余年历史。自古以来，香火鼎盛，是祭水防患的圣地。每逢初一、十五善男信女纷纷前来上香拜佛，为豫南地区著名古刹之一。

Dawang Temple

Located on Cunbei Street in Hekou Village, Zhugan Town of Luoshan County, the temple is also known as Jinlong Temple ("Jinlong" literally means golden dragon). Dating back 800 years, it has long been popular as a place of worship, and it is seen as a sacred place to worship the god of water and to prevent floods.

三义观

建于清康熙四十年（1701 年），位于潢川县南城的南海湖北岸。该观坐北面南，总占地 4000 多平方米。观内大殿供奉“桃园三结义”的刘备、关羽、张飞的塑像。观内大殿前有一树铸造于清嘉庆十四年（1809 年）的铁旗杆，气势恢宏。

Sanyi Taoist Temple

Located on the west bank of Nanhai Lake in south Huangchuan County, the temple was built in 1701. This south-facing building has a main hall for locals to worship Liu Bei, Guan Yu, and Zhang Fei, who were heroes during the Three Kingdoms Period (220-280).

紫水塔

位于光山县东门外，因临紫水河而得名。始建于明末，被毁后于清光绪年间（1875 ~ 1908 年）重修至第六层。新中国成立后复原七层和塔刹。为八角七级楼阁式砖塔，通高 27 米。它融南塔的清秀挺拔与北塔的庄重劲秀于一体。是古老的光山县的一个重要标志，人们常用“紫水弦山”来象征光山县。

Zishui Pagoda

Zishui Pagoda is located by the Zishui River outside the east gate of Guangshan County. First constructed at the end of the Ming Dynasty (1368-1644), the building was rebuilt during the reign of Emperor Guangxu (1875-1908) of the Qing Dynasty (1644-1911). It is a seven-floor octagonal pagoda with a height of 27 meters, and its style combines the delicacy of pagodas in southern China and the grandeur of pagodas in northern China.

法眼寺

位于商城县黄柏山国家森林公园内，距县城 60 千米，南邻湖北，东接安徽。为明代高僧无念禅师于明万历二十八年（1600 年）所创建，闻名于鄂豫皖三省，有“楚豫禅宗”之说。

Fayan Temple

The temple is located on Huangbai Mountain in Shangcheng County and is 60 meters away from the county town. It neighbors Hubei Province to the south and Anhui Province to the east. In the year 1600 during the Ming Dynasty (1368-1644), an eminent monk, Wunian, led the construction of the temple.

王母观

位于光山县之南，新县之北，为界山。兼大别山之雄峻与淮水之清纯，极具清灵之气。登高一望，四野旷旷，唯王母观一峰独秀，峻伟雄奇，气夺豫楚。遍视山中，植被繁茂、珍稀动植物丰富。和风习习，薄雾轻绕，实人间仙境也。

Wangmu Taoist Temple

The temple is located on the border between Guangshan County and Xin County. It stands high on a mountain where visitors can have a nice panoramic view of the magnificent mountain ranges, rare animals, and lush vegetation.

净居寺

位于光山县西南大苏山腹地，有 1000 多年历史，是中国八大佛教宗派中开宗立派最早的天台宗的发祥地、始祖庭。寺处素有“九龙捧圣，四水归池”之奇妙景观中。寺中尚存有明万历“皇帝敕粉”碑、清康熙“钦赐大苏山梵天寺重建记”碑和“宋苏轼游净居寺诗并叙”碑等诗赋碑刻 30 余块。

Jingju Temple

Located in the heart of Dasu Mountain in southeastern Guangshan County, the temple has a history of over 1,000 years. It is known as the birthplace of the Tiantai sect of Buddhism. It is surrounded by green mountains and clear waters.

铁佛寺

位于商城县城关镇东南，寺体为木制结构，几经战火焚毁，于清光绪十八年（1892 年）由桂月和尚重建。因寺内曾存放过一尊 1.6 米高、重 5000 公斤左右的生铁铸造的铁佛而被成为“铁佛寺”。

Tiefo Temple

Located in the southeast of Chengguan Town of Shangcheng County, the temple is a wooden building first built in 1892, and it has been damaged several times during various wars. A 5,000 kilogram Buddha made of pig iron was once preserved here. That is why the temple was named Tiefo, which means “iron Buddha” in Chinese.

聚贤祠

位于信阳市贤山之巅，古时曾有贤士隐逸于此，所以贤山又名贤隐山。后修建此祠，因信阳历代乡贤、名宦共聚于此，故名聚贤祠。整个建筑占地1000多平方米，布局类似于四合院，分正屋和厢房。此处登高望远视野极佳，绝世风光。

Juxian Ancestral Temple

Located on the top of Xian Mountain in Xinyang City, the temple is also known as Xianyin Mountain. As many sages and celebrities used to gather here, it was named Juxian, which literally means "the gathering of wise persons." Its structure is very much like that of a rectangular courtyard. Visitors can have a bird's eye view of the surrounding area from the temple.

息县谯楼

位于息县县城，始建于元元贞元年（1295 年），后为战火所毁。明、清两代进行多次重建和整修，1984 年，息县人民政府仿古重修。全楼三楹，回廊四面，丹栋青椽，琉瓦镶嵌，檐牙高翘，殿脊走兽，前装花门掩映，后壁菱窗透空，中间拱门为古州衙之大门。

Watchtower in Xi County

Located in Xi County, the watchtower was rebuilt during the Ming (1368-1644) and Qing dynasties (1644-1911) and revamped in 1984 (it was originally completed in 1295). It consists of three rooms and four corridors on each side, and it is decorated with red pillars, gray beams, glazed roof tiles, eaves pointing to the sky, and tiny statues of animals on the ridges.

三里店遗址

位于信阳市南郊三里店西靶场，面积 42 万平方米，文化层厚三米。出土陶器以红色泥质和黑色磨光陶为主，器形有鼎、豆、鬲、碗、盆、罐、瓮等，多饰绳纹、篮纹，石器有斧、凿等。属仰韶文化晚期、龙山文化早期、以及商、周文化的遗存。

Sanlidian

Located in Sanlidian at the southern suburb of Xinyang City, the relic is a cultural heirloom dating back to 5,000 BC - 3000 BC. Many daily necessities and objects, like pottery products and stone vessels, have been unearthed here.

沙冢遗址

位于淮滨县城西北肖南庄，面积约1600平方米，文化层厚一至三米。1979年试掘，出土有釜形鼎、钵、杯、罐、豆、盘等陶器，陶质以加砂土灰陶和泥质黑陶为主。石器有斧、镰、锛、刀、锥、镞等，还有骨器和大量的蚌壳、鹿角、兽骨等。属龙山文化遗存。

Shazhong

The relic is a cultural artifact of the Longshan Culture (2500 BC-2000 BC) and is located in Xiaonanzhuang Village northwest of Huaibin County. Many ceramic products, stone vessels, bone objects, clam shells, antlers, and animal bones have been unearthed here.

期思台地遗址

位于淮滨县城东南期思村西北，东临死河，南部紧靠期思古城。遗址为高出地面四至六米的台地，台地南高北低，东、西、北三面为陡坡，文化层厚约四米。该遗址属新石器时代晚期遗址。

Qisi Mesa

This site is a relic of the late Neolithic Age and is located in southeastern Huaibin County. The mesa is 4-6 meters high with steep slopes on three sides and a cultural layer of 4 meters.

高台庙遗址

位于淮滨县城西北防胡镇冯庄村黎庄东。遗址为灰土堆积，中部隆起，高出地面二至五米。范围东西长 100 米，南北宽 98 米，文化层厚约三米。属新石器时代晚期遗址。

Gaotai Temple

The site is a relic of the late Neolithic Age and is located in Fengzhuang Village of Huaibin County. Now, only lime soil is left. The left temple is 2-5 meters above the ground with a cultural layer of 3 meters deep.

南山嘴文化遗址

位于信阳市平西村。考古队曾在此处挖掘有五座春秋早期墓葬。尤其是樊君夫妇合葬墓的出土，对研究春秋战国的丧葬礼俗提供了珍贵史料。文化层堆积较厚，跨越时代较长，为研究淮河上游地区新石器时代文化的发展概况提供了重要依据。

Cultural Relic of Nanshanzui

The site is located in Pingxi Village, Wuxing Town of Xinyang City. Around the relic, five tombs of the early Spring and Autumn Period (770BC-476BC) were unearthed, which are of great significance to the study of burial customs during the Spring and Autumn Period (770 BC-476 BC), as well as the Warring State Periods (475 BC-221 BC).

城阳城址

位于信阳市北，为古楚国国都，已有2700多年历史。城阳城保护区内现有大小楚墓一百多座，现已发掘的有八座楚墓，共出土各类珍贵文物两千多件。其中1957年发掘的一号墓震惊全国，出土了我国第一套完整的青铜编钟，用其演奏的“东方红”乐曲随着我国第一颗人造地球卫星回响太空。

Chengyang

Located in the north of Xinyang city, the relic was once the site of the capital city of the Ancient Chu Kingdom. With a history of 2,700 years, the site boasts over 100 tombs from the Chu Kingdom and more than 2,000 pieces of cultural artifacts have been unearthed here.

蒋姓故国遗址

位于今淮滨县期思镇，属于西周时期周公在淮滨县一带的诸侯国之一，蒋国是周公姬旦的第三子伯龄于公元前 11 世纪左右建立的，周顷王二年（前 617 年）左右被楚国所灭，其后代以国名为姓。城遗址呈长方形，东西长 1700 米，南北长 500 米，城墙高 9~12 米。

Jiang Kingdom

Located in Qisi Town, Huaibin County, this is where the Jiang Kingdom was founded during the Western Zhou Dynasty from 1040 BC to 617 BC. Later, the kingdom name was adopted as a family name. The remains consist of a rectangular building that is 1,700 meters long, 500 meters wide, and has walls of 9 to 12 meters high around it.

琵琶台

位于信阳市浉河区，前身是信阳古八景之一的长台古渡，古称“长台渡”，是明清时代重要的淮河渡口，为盐、粮及其他物质的集散地。

Pipatai Ferry

Located in Shihe District in Xinyang city, the site is known as Changtai Ferry. As an important ferry on the Huaihe River during the Ming (1368-1644) and Qing dynasties (1644-1911), it was a hub for salt and grain transportation.

司马光故居

位于光山县城正大街，是北宋著名政治家、史学家、文学家司马光的出生地。为二进四合院落，大门门屋北檐外有照壁，有前厅、厢房、书斋、后堂等，均为悬山式砖木结构建筑。院中植柏树、胡桃、梧桐，中置司马井。

Former Residence of Sima Guang

Located in the middle section of Zhengda Street in Guangshan County, the residence was the birthplace of Si Maguang, a famous historian and writer during the Northern Song Dynasty (960-1127). It is a rectangular courtyard consisting of two rows of rooms. There are a front hall, living rooms, a study, a back hall, and a Sima well.

张云墓

位于信阳市何寨彭家湾，张云，生卒年不详，曾任朝廷要职。其墓系张云夫妇合葬。

Tomb of Zhang Yun

The tomb is located in Hezhai Village of Pengjiawan Town in Xinyang City. Zhang Yun was a government official of the Qing Dynasty (1644-1911) and his dates of birth and death are not clear.

黎世序墓

位于罗山县城刘店村。黎世序（1772～1824年），字景和，号湛溪，初名承惠，清嘉庆元年（1796年）进士。曾任江西星子知县、南昌知县、镇江知府、淮海道员、南河河道总督。墓葬在一山坡上，坐北朝南，直径约11米，下部为青砖垒砌，上部堆有封土。西北距御碑亭200米。

Tomb of Li Shixu

The tomb is located in Liudian Village, Dingyuan Township, to the south of Luoshan County. Li Shixu was a government official in the Qing Dynasty (1644-1911). The south-facing building was made of gray bricks on the bottom and a sand mound on the top.

鄂豫皖革命纪念馆

位于信阳市，占地三万平方米，2007 年 4 月 28 日建成开馆。纪念馆按时代顺序陈列展览，以大量文字图片与实物对应的方式，全面展示鄂豫皖革命根据地形成、发展和不断壮大的过程，以及从 20 世纪 20 年代至 1949 年这里发生的重大历史事件。

Memorial Museum for the Revolution in Hubei, Henan and Anhui

Located in the city of Xinyang, the museum was completed in 2007. A vast amount of paper materials, pictures, and other objects are on display here so that visitors can have a full grasp of the history of the revolutionary base.

邓颖超祖居

位于光山县司马光中路白云巷内，是原全国政协主席、中国妇女运动的先驱邓颖超同志的祖父及父亲居住的地方。该建筑占地2000平方米，坐北朝南，前后为两个独立四合院落，现存清代建筑房屋30多间。建筑结构严谨，格扇门窗古朴典雅，是一座典型的具有南方特点的清代建筑。

Ancestral Residence of Deng Yingchao

Located in Baiyun Alley, Middle Sima Guang Road, this site was the residence of Deng Yingchao's father and grandfather. There are two separate rectangular courtyards and over 30 houses with the distinctive features of the Qing Dynasty (1644-1911). Deng Yingchao (1904-1992) was the wife of Zhou Enlai, the first premier of the People's Republic of China and she then became a pioneer of feminist movements in China.

万海峰将军旧居

位于光山县的泼河镇小蔡湾、槐店小黑湾。万海峰将军生于 1920 年，幼时家境贫寒，1933 年参加红军，1937 年加入中国共产党，参加了长征，历经抗日战争、解放战争、抗美援朝，1988 年被授予上将军衔。

Former Residence of General Wan Haifeng

The residence was located in Pohe Town of Guanshan County. Born in 1920, General Wan Haifeng participated in the Anti-Japanese War (1937-1945), the War of Liberation (1946-1949), and the Korean War against the invasion of the U.S. (1950-1953). He was given the rank of general in 1988.

许世友将军旧居

位于新县田铺乡河铺村，坐北朝南，占地322平方米。“吞”字大门，上方悬挂着“许世友将军故居”的匾额。大门后为正房和会客厅，系后来修建，客厅里有将军工作和生活片段陈列，后面与客厅相连有四间厢房。东首第一间曾是将军的卧室，里面有他当年结婚时的老式木床。许世友，为著名爱国将领，在抗日战争中等有过突出贡献，1955年被授予上将军衔。

Former Residence of General Xu Shiyou

The residence is located in Xujiawa, Hepu Village, Tianpu Town of Xin County. The south-facing building has a grand gate in the back, a main hall, and a parlor. The old wooden bed used when the general got married is still in the living room. Xu Shiyou (1905-1985) served his position with distinguished excellence during the Anti-Japanese War and the War of Liberation, and he received the rank of general in 1955.

秦树声旧居

位于固始县乐道村。秦树声是清末民初著名书法家，曾两中进士，清任工部主事等官职，充任《地理钧稽图志》总纂，民国时为《地理志》总纂，其墨迹收入《民国书法》，他在全国最先引进两部印刷机。故居占地 6000 多平方米，现存堂楼上下两层 28 间。

Former Residence of Qing Shusheng

Located in Ledao Village, Duanji Town of Gushi County, the residence was once the home of Qing Shusheng, a famous calligrapher in the late Qing early Ming dynasties (1840S~1920S). His calligraphic works were collected in the album Calligraphy during the Period of the Republic of China. The two-story house has 28 rooms all together.

郑维山将军故居

位于新县泗店乡泗店村屋脊洼，故居保存完好，为大别山区常见青砖瓦房。现在已辟为郑维山将军生平事迹陈列馆，进门正中是郑维山将军的半身雕像。郑维山，河南省新县人，为抗日将领，1995 年被授予中将。

Former Residence of Zheng Weishan

The residence is located in Wujiwa, Sidian Village of Sidian Township of Xin County. It is a well-preserved building made by bricks and covered with tiles, which are commonly used construction materials in the Dabieshan region. Zheng Weishan (1915-2000) was a senior military officer that made great achievements during the Anti-Japanese War and China's War of Liberation.

高敬亭将军旧居

位于新县新集镇境内，始建于清咸丰十年（1860 年），后被毁，于 1982 年重新整修。旧居为三间青砖瓦房，内设高敬亭遗像和经后人搜集的几张战场照片。2005 年，县文物局再次对故居进行了整修，建成故居正屋六间，包括生平事迹简介陈列室、图片展厅、将军雕像室、院墙及门楼。

Former Residence of Gao Jingting

Built in 1860, the residence is located in Xinji Town of Xin County; it is made of bricks and covered with tiles. Gao Jingting (1907-1939), one of the founders of the Twenty-Eighth Red Army, is known for his great achievements on the battlefield.

吉鸿昌将军旧居

位于光山县城西胡围孜村，原为裴氏祠堂，始建于1919年。建筑坐北面南，二进18间，占地约640平方米。该建筑为一天井院落，内有回廊一周，有正殿和东西厢房。吉鸿昌为河南省扶沟人，抗日爱国将领。1931年春和1932年冬，吉鸿昌将军两次率部来光山，居住在此。

Former Residence of Ji Hongchang

Located in Weizi Village, Chengguan Town, in western Guangshan County, the residence was built in 1919 and was once used as an ancestral hall for the Pei family. Within the south-facing courtyard are corridors, a front hall, and two bedrooms at the east and west ends of the courtyard. Ji Hongchang (1895-1934) was a famous general during the Anti-Japanese War.

汪厚之旧居

位于潢川县，汪厚之，曾任中共河南省省委委员、豫东南特委书记，豫东南早期党的领导人和活动家。

Former Residence of Wang Houzhi

The residence is located in Huangchuan County. Wang Houzhi (1900-1928) was one of the communist leaders and activists in the early period of the revolution in the southeast of Henan Province.

熊少山旧居

位于光山县马畈镇柳林村。熊少山，生于清光绪十年（1884 年），为早期社会革命活动家。1930 年 11 月，熊少山在卡房古店潭洼竹园祠堂的群众大会上慷慨激昂演讲时，突然病逝讲台上，以身殉职。

Former Residence of Xiong Shaoshan

The residence is located in Guangshan County. Xiong Shaoshan (1884-1930) was a communist activist during the early years of the revolution.

杜彦威故居

位于光山县殷鹏乡李庄村，杜彦威，1924 年加入中国共产党，曾与熊少山等领导殷鹏区农民运动，1928 年加入红军，1931 年在新县卡房乡与敌作战时牺牲。旧居为五间硬山灰瓦房，面积 100 多平方米。1929 年春，殷鹏区农民起义司令部曾秘密设立于此。

Former Residence of Du Yanwei

The residence is located in Lizhuang Village, Yinpeng Township of Guangshan County and consists of five houses with flush gable roofs and covered with gray tiles. Du Yanwei (1908-1931) was a revolutionary martyr who joined the Red Army in 1928.

胡煦故居

位于光山县弦山胡围孜村。胡煦，字沧晓，号紫弦，生于清顺治十二年（1655年）。出身于书香世家，历任翰林院检讨等职，官至礼部左侍郎（掌管全国典礼、科举和学校等事务的副职），是历经清康熙、雍正、乾隆三朝的老臣。精通文、史，擅长书画。

Former Residence of Hu Xu

The residence is located in Weizi Village of Guangshan County. Hu Xu (1655-1736) was a senior official during the reigns of Emperor Kangxi (1662-1723), Emperor Yongzheng (1723-1736) and Emperor Qianlong (1736-1796). He was also a master in literature, history, calligraphy and painting.

白雀园革命旧址

位于光山县白雀镇白雀街，包括“白雀园大肃反”监狱旧址、白雀园苏维埃政府、鄂豫皖省政府治安保卫局分局旧址、明代城门及古城墙、红军井等。是整个鄂豫皖革命根据地旧址群的重要组成部分。旧址多为清和民国时期建筑，共有房屋 21 间，建筑面积 500 多平方米。

Revolution Relic of Baique Park

The relic is located on Baique Street, Baique Town of Guangshan County. The site was mostly built during the Qing Dynasty (1644-1911) and the Republic of China Period (1912-1949) with Kuo Min Tang as the ruling party. There are 21 houses and the area contains important historical sites of the Revolution in Hubei, Henan and Anhui against Kuo Min Tang.

花旗楼

位于信阳鸡公山南岗狼牙令南侧 135 号楼，始建于 1918 年，由英国汇丰银行老板潘尔恩所建。前临悬崖，后靠峭壁，为两层西式楼房，建筑面积 172 平方米。1937 年、1938 年蒋介石夫妇两次来鸡公山均住于此。楼西有 1937 年修的 70 米长防空洞，洞口与楼底相通。

Huaqi Building

Located on the southern side of Jigong Mountain, the two-story, Western-style building was built in 1918 by the British. There are cliffs on both sides of the building.

姊妹楼

位于信阳鸡公山南街南头西山岗，建于 1912 年。北楼系袁世凯侄孙袁英所建，南楼系南阳镇守使吴庆桐所建。这两栋二层别墅建筑造型相似，面阔 17 米多，进深 14 米。墙体以花岗岩筑砌，上为坡形红瓦顶。门庭两侧有前廊，上下施檐柱栏杆。楼外侧设楼梯道。

Zimei Building

Located in West Hill on the south of South Street on Jigong Mountain of Xinyang County, the two-story villa was built in 1912 with walls made of granite slabs and red tiled roofs.

瑞典式大楼

位于信阳鸡公山北岗东侧，建于 1926 年。原为瑞典传教士兴办的瑞华学校，1926 年改建为三层石、砖、木混合结构楼房。平面长方形，中楼用券式门窗，两翼方门窗，上为人字形大坡铁皮瓦顶，楼下还有地下室，保存完好。

Swedish-Style Building

It was built in 1926 to the east of the northern side of Jigong Mountain in Xinyang City. The building was once a Swedish-Chinese School constructed by Swedish missionaries. In 1926, it was rebuilt into a three-story building made of stones, bricks and wood.

颐庐

位于信阳鸡公山中心区南北街中间山头上，建于 1921 年。系吴佩孚部下勒云鄂之别墅，由中国人设计和建造的中西合璧的建筑。规模和气势均压倒当地的洋房，故又称“志气楼”“正气楼”。面积 1274 平方米，通高 21 米。坐落在台基上，顶部有六角形的翘檐小亭，张学良在此办过东北中学。

Yilu Pavilion

Located on the edge of a cliff between North and South streets in central Jigongshan, the pavilion was designed and built by the Chinese in 1921. It combines Eastern and Western styles and covers an area of 1,274 square meters. It is a 21-meter hexagonal building with eaves pointing to the sky.

小教堂

位于信阳鸡公山北岗宝剑泉西南公路旁，又称“小礼拜堂”。清光绪三十三年（1907 年）由美国人兴建，1913 年扩建。建筑面积 222 平方米，西式建筑，放坡甚陡，大门上部筑有十字架，门窗为尖拱形，为当地各国教徒所共用的教堂。

Small Church

Located near the southwest highway of Baojian Spring on the north side of Jigong Mountain, the church was built by Americans in 1907. It covers an area of 222 square meters and there is a cross right above the door.

望河楼

位于潢川县小潢河北岸，1932 年由张钫先生倡导，在“古今庙”的旧址上建立而成。楼原高五层，约 26 米，现仅存三层。建筑面积 147 平方米，集楼、台、亭、阁于一体。第一层有碑记，第二层的楼外四面刻有文字，西、南、北面分别是“汲古”“滴翠”“崇文”，东面缺失。

Wanghe Building

Located on the north bank of Xiaohuang River in Huangchuan County, the building was built in 1932. It used to be a 5-story building with a height of about 26 meters, but only 3 floors remain today. There are memorial steles on the first floor, and on the walls of the second floor are engraved decorations.

观音山

位于商城县城北的河凤桥乡境内。山上建有规模较大的道教庙宇——云极观，是信阳市道教协会所在地，也是豫南山区有名的道教胜地。每年庙会期间，热闹非凡。

Guanyin Mountain

Located in Hefengqiao Town to the north of Shangcheng County is the Yunji Taoist Temple located on Guanyin Mountain.

西九华山

位于固始县东南境，属大别山脉中段，观赏面积80平方千米，是中原地区最大的集“茶、竹、禅、山水情”为一体的生态旅游胜地。

Western Jiuhua Mountain

Located in the southeast of Gushi County, the mountain is in the middle of the Dabieshan Mountain Range. It is the largest tourist site in middle China boasting beautiful scenery, tea, bamboo and Buddhism.

信阳毛尖

中国十大名茶之一，河南省著名特产。主要产地在信阳市和新县、商城县及境内大别山一带，在唐代时即有记载，清代已成为全国名茶之一，有“绿茶之王”之美誉。1915 年在巴拿马万国博览会上与贵州茅台同获金质奖。

Xinyang Maojian Tea

As one of the top ten teas in China, Xinyang Maojian tea is known as the “King of Green Teas.” It is mainly produced in downtown Xinyang City, Xin County, Shangcheng County, and in the Dabieshan Mountains.

信阳茶神节

茶神陆羽多次考察淮南茶区，在紫阳洞中写出《茶经》一书。信阳人为纪念其对信阳茶的研究和贡献，特定每年清明节为信阳茶神节。茶神节当天，茶农聚集，进行祈愿茶丰收的茶神祭、采茶和加工茶叶比赛、茶艺表演等活动，同时参拜陆羽的神位。

Festival to Worship the God of Tea in Xinyang

Lu Yu of the Tang Dynasty (618-907) was revered as the "God of Tea" for his great contribution to tea culture. Lu's book, *The Classic of Tea*, marks the first monograph about tea culture in the world. In honor of Lu Yu, April 5th was set as the festival to worship the "God of Tea" by people in Xinyang.

罗山皮影

是一种古老的、具有独特魅力的民间艺术。始于明代，起源于河北滦州，从明嘉靖年间（1522 ~ 1566 年）开始在罗山县繁衍生长。经过数代皮影艺人的改进、提高，已达到较高水平。影人制作美轮美奂，栩栩如生。生、旦、净、丑行当齐全，音乐旋律流畅，唱词、道白雅俗共赏，是中国戏剧园中的一块瑰宝。

Shadow Play in Luoshan County

Shadow play is an ancient and unique folk art. It first started in Luanzhou, Hebei Province during the Ming Dynasty (1368-1644) and Luoshan County has a history of more than 400 years performing it.

信阳民歌

信阳市素称河南的歌舞之乡，在民间音乐、舞蹈等传统文化方面有着丰厚的存量，在风格上异于淮河以北的河南省内各地区。经过多次的变化，纯原始的汉族音乐舞蹈已不多见，代之以原民间音乐舞蹈素材为基础，再创作而形成的一批音乐、舞蹈节目。

Xinyang Folk Song

The city of Xinyang has long been known as the "Town of Singing and Dancing" in Henan Province. Folk songs in Xinyang have incorporated different styles from those of other areas in Henan. Many existing songs are mostly based on local music and dances from the past.

商城花伞舞

花伞舞是一种民间舞蹈，商城县素称花伞舞之乡。起源于唐宋时期，代表着大别山民间舞蹈的风格，以一人或多人形式表演，演员们表演扛伞、转伞、甩手帕、挽花等动作，美轮美奂。随着时代的发展融入很多新的元素和内容，是当地民间活动的主要表演内容。

Dancing with Colorful Umbrellas in Shangcheng County

This local dance dates back over 1,000 years ago. Performers will act as if they are carrying or spinning umbrellas while tossing handkerchiefs and displaying "wanhua" (a movement of Taiji Boxing).

潢川观音庙会

佛教于唐末宋初传入潢川县，明、清为发展鼎时期。在当时州治的十多座寺庙中，观音庙会在潢川南海观音禅寺。农历二月十九、六月十九、九月十九，相传是观音出生、出家、得道的日期，每年此时佛教信众云集南海观音禅寺，方圆百里的香客赶来朝拜，形成小南海观音庙会。

Guanyin Temple Fair of Huangchuan County

The temple fair is hosted in the Guanyin Buddhist Temple in Huangchuan. On the 19th of the second month, the 19th of the sixth month, and the 19th of the ninth month by the lunar calendar, followers of Buddhism will gather together at the temple to pay their respects and worship because the 3 days are respectively regarded as the date when Avalokitesvara was born, the day he became a Buddhist monk, and the day he became a god.

肖营泥塑

淮滨县肖营村传统泥塑，采用当地粘土，经过仔细筛选、和泥、捏制、烧炼、彩绘等工序制作而成，多制作动物和人物造型。彩绘是以黑色、棕色打底，再描绘上白土粉、大红、大绿、大蓝、大黄等条纹。肖营泥塑以小手工作坊生产为主，农闲时节，全家男女老少围坐在一起，边制作边说笑，十分具有生活气息。

Xiaoying Clay Sculpture

Xiaoying clay sculptures are a traditional hand-made craft in Xiaoying Village of Huaibin County. Sculptures of animals and human beings are the most common creations. These lifelike sculptures are made with bright colors. Craftsmen use local clay and go through numerous procedures for making the sculptures.

信阳地锅饭

信阳地锅饭是结合信阳特殊地理位置和环境下产生出的做饭方式。是用黄泥和砖块砌成的几何形状的锅台，前边有开口用来添柴火，上面留出大的圆形放上铁锅，在出口上方砌上烟囱用来排烟，锅里煮饭，锅下添柴。

Meals cooked in ground kettles in Xinyang

The top of a kitchen range made in this style is constructed of clay and bricks, with an opening in the front for adding firewood. There is also a chimney to contain the smoke. A large iron pan is used for cooking.